ECONOMIC ZOOLOGY : A HANDBOOK PART I

REVATHY S

Copyright © 2016 Revathy S

Published by Witness

Printed by CreateSpace, United States of America

All rights reserved.

ISBN-13: 978-1539527473
ISBN-10: 1539527476

PREFACE

This book contains a comprehensive account on various aspects of Economic Zoology. Module I Aquaculture cover General Identification-Morphology, Feeding habit and Economic Importance of certain selected cultivable aquaculture species and Module II Apiculture highlights the species of honey bees used in Apiculture, characteristic features of queen, worker and drone, bee keeping equipment, by products of apiculture industry and simple methods for determining the purity of honey.

The book is written in a very simple and lucid language enabling both the students of Zoology as well as readers without biology background to understand the subject easily.

As in the name of the book **"ECONOMIC ZOOLOGY :A HANDBOOK - PART I "**, the book has very easy notes enabling the students to easily recollect and prepare for exams.

This book **"ECONOMIC ZOOLOGY : A HANDBOOK - PART I"** is a very first attempt and suggestions are welcome and will be incorporated in the subsequent editions.

Author

CONTENTS

1. Aquaculture

    a. *Catla catla*
    b. *Labeo rohita*
    c. *Cirrhinus mrigala*
    d. *Ctenopharyngodon idella*
    e. *Cyprinus carpio*
    f. *Etroplus suratensis*
    g. *Tilapia mossambica*
    h. *Penaeus indicus & P.monodon*
    i. *Perna viridis & P.indicus*
    j. *Pinctada fucata*

2. Apiculture

    a. Species of Honey Bees
    b. Characteristic features of Queen, Worker and Drone
    c. Bee keeping Equipments
    d. By products of Apiculture Industry
    e. Simple methods for determining the purity of honey

ACKNOWLEDGMENTS

I would like to sincerely acknowledge my teachers, colleagues, students, friends and family members for their constant support and encouragement.

I. AQUACULTURE

Aquaculture augments food production and a number of species are now cultured in different parts of the world. Each species has characteristic identifying features, feeding habits and economic importance. This module covers General Identification-Morphology, Feeding habit and Economic Importance of certain selected cultivable aquaculture species.

- *Catla catla*
- *Labeo rohita*
- *Cirrhinus mrigala*
- *Ctenopharyngodon idella*
- *Cyprinus carpio*
- *Etroplus suratensis*
- *Tilapia mossambica*
- *Penaeus indicus & P.monodon*
- *Perna viridis & P.indicus*
- *Pinctada fucata*

Catla catla

Common name: Catla

General Identification-Morphology & Feeding habit

- Short and deep body, somewhat laterally compressed with depth more than head length
- Very large head
- Body with conspicuous large cycloid scales
- Head lack of scales
- Snout blunt and rounded
- Wide and upturned mouth with prominent protruding lower jaw
- Upper lip is absent
- Lower lip is very thick
- Barbels are absent

- Dorsal fin is inserted slightly in advance of pelvic fins, shorter anal fin pectoral fins long extending to pelvic fins; caudal fin forked; lateral line is with 40 to 43 scales.
- Colour is greyish on back and flanks, silvery-white below; fins dusky.
- They are surface and side vegetation feeders.

Economic importance

It is a valuable food fish cultured in fresh water ponds. The fish is fleshy & renowned for its delicacy & valued very high in the market.

It is highly growing species which can weigh up to 4 kg in the first year itself, provided they are properly fed. It is a very delicious food and supply huge amount of protein for people. So, its demand is extensive.

Labeo rohita

Common name: Rohu

General Identification-Morphology & Feeding habit

- Body is laterally compressed & fusiform, attaining maximum length of one meter.
- Colour blackish grey on the back & silvery white below.
- Body is covered with cycloid scales which are overlapping.
- Head is prominent with blunt snout.
- Eyes are large without eyelids.
- Mouth is sub-terminal, directed downwards & surrounded by thick lips.
- Upper lip with a pair of short barbells & lower lip fringed.
- Jaws are without teeth.
- Dorsal fin is large and seen about the centre of the body.
- Pectoral fins without spines.
- Tail is homoceral and small.
- Rohu is a column feeder and feeds on phyto and zooplanktons.

Special Features: Both upper & lower lip have an inferior transverse fold, which is fringed on the lower lip.

Economic importance

It is a most popular food fish. Its flesh is delicious and relished very much for the taste. The flesh is rich in protein content. The fish is mostly cultivated with catla in fresh water ponds & lakes.

Interspecific, intergeneric hybrids and variants of rohu have been produced. The most promising intergeneric hybrid is with male Catla & female Rohu. The hybrid has the quick growth of Catla and small head characteristic to Rohu (Talwar and Jhingran, 1991).

Cirrhinus mrigala

Common name: Mrigal

General Identification-Morphology & Feeding habit

- Bilaterally symmetrical and streamlined body
- Body is with cycloid scales
- Head is devoid of scales
- Snout is blunt
- Mouth is broad and transverse
- Upper lip is not continuous with lower lip
- A pair of short rostral barbels present
- Dorsal fin is as high as body and has 12 or 13 branched rays
- Pectoral fins shorter than head; caudal fin is deeply forked
- Lateral line has 40-45 scales
- The colour of the fish is dark grey above, silvery beneath; dorsal fin greyish; pectoral, pelvic and during breeding season the anal fins become orange-tipped.

- It is a bottom feeder and feeds on small insects, organic and decaying particles from the bottom waters or inside the soil.

Economic importance

It is popular as a food fish cultured in freshwater. They are very fast growing and tasty. The species is highly suitable for polyculture along with Catla and Rohu. This fish has a rapid growth rate.

Ctenopharyngodon idella

Common name: Grass Carp

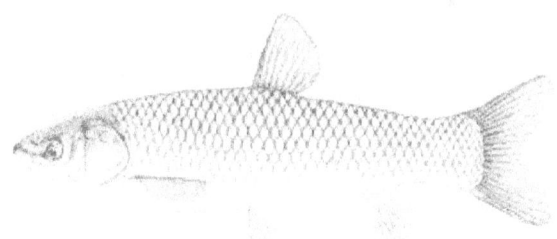

General Identification-Morphology & Feeding habit

- The body is elongated, chubby, torpedo-shaped.
- The mouth is terminal and oblique slightly.
- Lips are firm and without barbels.
- The lateral line contains 40 to 42 scales.
- The dorsal fin has 8 - 10 soft rays, and the anal fin is set closer to the tail.
- Body is dark olive, shading to brownish-yellow on the sides, with a white belly.
- Grass carp feeds on aquatic herbs and grass like plants. It also cleans the water unhealthy aquatic vegetation and hence used in aquatic weed control.

Economic importance

Grass carp is a cultivable and highly growing species. It can weigh about 4.5 kg in the first year if feed is provided properly. The fish is very tasty and serve as a protein source for humans.

Cyprinus carpio

Common name: Common Carp

General Identification-Morphology & Feeding habit

- It is characterized by the deep elongated somewhat compressed body and serrated dorsal spine.
- The mouth is terminal foradult and sub-terminal for young fishes.
- Lips are thick.
- Two pairs of barbels are present, shorter ones on the upper lip.
- Dorsal fin with 17-22 branched rays and a strong, toothed spine in front; dorsal fin outline concave anteriorly. Posterior edge of 3rd dorsal and anal fin has spines with sharp spinules.
- Lateral line has 32 to 38 scales.

- Colour is variable. It may be brownish-green on the upper and back sides, with shades of golden yellow ventrally. The fins are dusky has a reddish tinge ventrally.
- They feed on macrophytes in water.

Economic importance

Common carp can be cultured in small water bodies and ponds. It is well adaptable to the varying environments. It is well suited for extreme climates and hence a hardy and tolerant species. It has good market value with delicious taste.

Etroplus suratensis

Common name: Pearl spot

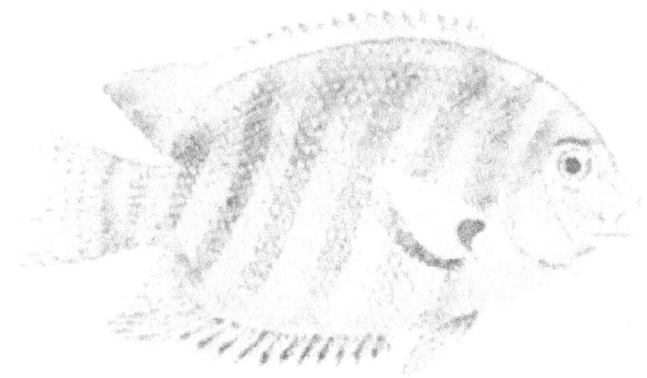

General Identification-Morphology & Feeding habit

- Body is deep and laterally compressed.
- Mouth is small and terminal.
- Lips are thin and jaws are equal with teeth.
- Above the pectoral fin base is the dorsal fin; Emarginated caudal fin is present; pelvic fin has one spine.
- Scales are ctenoid.
- Lateral line is interrupted.
- Body is light greenish with eight yellowish oblique bands.
- Pearly spot is present on the scales above lateral line.
- It feeds on algae, plant material and insects.

Economic importance

Pearl spot is a highly nutritive food with good meat content. It has delicious taste. It is candidate species for aquaculture in ponds, brackish and fresh waters. This fish is expensive, highly priced and is available throughout the year. It is known for its compatibility to live together with a wide variety of fishes. It has high growth rate. On account of their characteristic coloration and banding patterns, they are also valued as ornamental fishes.

Tilapia mossambica

Common name: Tilapia

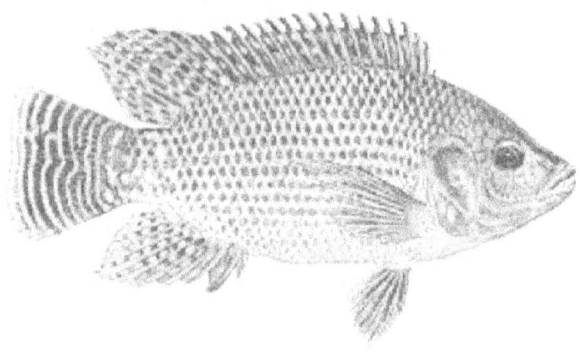

General Identification-Morphology & Feeding habit

- The body is laterally compressed.
- They have an interrupted lateral line which is a distinguishing feature Cichlid family of fishes.
- Mouth is large and oblique.
- Scales are cycloid
- Dorsal fins are long and heavy spines on the front portion. The pelvic and anal fins are with spines. The caudal fin is rounded.
- Coloration is grey to olive on top; dull yellow to grey-green on sides; below part is yellow. The dorsal fin is black with red border; the upper lip is blue.
- It is omnivorous feeder feeding on blue green algae, diatoms, macrophytes, green algae, detritus material,

copepods, cladocerans, semi digested matter and benthic algae, dinoflagellates, larvae of insects and bivalves.

Economic importance

Tilapia is a fish suitable for aquaculture. They are easily spawned. They feed on wide variety of feeds both natural and artificial. They can tolerate poor water quality, and grow rapidly. The species has high nutritional value and mild taste.

Penaeus monodon

Common name: Giant tiger prawn, Asian tiger shrimp

General Identification-Morphology

- The rostrum is well developed and with teeth.
- Carapace is without sutures.
- Basal spines on first and second pereiopods and exopods on the first to fourth periopods are usually present. The most distinct features for identification of this species is the absence of exopod in the fifth pereiopods.
- Telson lacks sub apical spines.
- Thelycum is closed in females.
- Male has symmetrical Petasma and thin median lobes.
- Body colour can vary from green, brown, red, grey, blue and there is transverse band colours on abdomen.

Penaeus indicus

Common name: Indian Prawn

General Identification-Morphology

- The rostrum slightly curved at tip and sigmoidal-shaped with teeth,
- Carapace is smooth.
- Telson is devoid of lateral spines.
- Petasma with disto-median projections strongly curved and overhanging distal margin of costae.

- Thelycum of females has 2 semi-circular lateral plates,
- Body is semi-translucent, somewhat yellowish white (small specimens) or greyish green and with dark brown dots.

Economic importance of Prawns

Food value

- ❖ serve as fresh food
- ❖ They can be made into dried or canned food
- ❖ Value-added products, microwavable or ready-to-cook tempura, sushi, spring roll and balls are also made.

Other values

Chitin is an N-Acetyl glucosamine polymer with monomers bond cellulose seen in the exoskeleton is used as

- a fertilizer for increasing crop yield
- a binder in dyes, fabrics, and adhesives
- reproducible form of biodegradable plastic
- Chitin's flexibility and strength make it favourable as surgical thread

Perna viridis

Common name: Asian Green mussel

General Identification-Morphology

- It is a large bivalve (> 80 mm).
- The shell is smooth and elongate shell typical.
- Concentric growth rings are present and a distinctly concave ventral margin on one side. The colour is characteristic green.
- The inner portions of the valves are smooth and shining blue to bluish-green colour.
- A pair of hinge teeth is also present.
- Presence of proteinaceous byssal threads.

Perna indica

Common name: Brown mussel

General Identification & Morphology

- Shells are thick, equivalve, elongate, triangularly ovate in outline.
- The posterior margin is rounded and the ventral margin is straight.
- The external colour is dark brown and the interior is shining.

Character	*Perna viridis*	*Perna indica*
External Colour	Brown	Green
Anterior end shape	Pointed & straight	Pointed & beak like
Size of hinge plate	Thick narrow	Thick broad
Mantle margin colour	Brown	Yellowish green
Dorsal ligamental margin	Straight	Curved
Ventral shell margin	Almost straight	Highly concave

Economic importance

They are used as food due to its fast growth and culture potentials. Mussels are exported to different countries in frozen or dried condition. Various value added products of mussels like seafood cocktails are prepared and marketed by seafood export firms in India.

Mussels also contain pharmacologically important ingredients for arthritis and sinusitis. It is well known that mussels accumulate large amount of zinc and has a number of health benefits, it is a proven immunity booster, promotes growth, mental alertness and aids in proper brain function. Mussel shells are used for landfill.

Mussel shells also are sources of calcium carbonate. The shells have been used as an animal feed additive, for the production of mortars, a liming agent and constituent in fertilizers.

Pinctada fucata

Common name: Pearl Oyester

General Identification-Morphology

- The hinge is long.
- Both valves have hinge teeth. The convexity of the valves is also greater
- The posterior ear is quite well developed. r than in other species of the genus.
- The shells are reddish brown or yellowish brown in colour.
- The non-nacreous border on the inner surface of the valves possesses brownish or reddish patches.
- The valves have well developed nacreous, golden yellow in colour and with a bright, metallic lustre.

Economic importance

Pearl-oyster shells and pearls have long been used as jewellery items by humans. The adductor muscle is used as food in different regions. The shell is turned into complex furniture and other decorative items. The pearl oyster shell also has many health benefits. Historically, the largest use of the pearl resource was for mother of pearl also called as nacre. This shell material was used for making buttons. Shells also are sources of calcium carbonate.

II. APICULTURE

Apiculture is the art of rearing and keeping honey bees. This module highlights the species of honeys bees, Characteristics of Queen, Worker and Drone, Bee keeping Equipments and By products of bee keeping.

- Species of Honey Bees
- Characteristic features of Queen, Worker and Drone
- Bee keeping Equipments
- By products of Apiculture Industry
- Simple methods for determining the purity of honey

Species of Honey Bees

	Apis dorsata	*Apis indica*	*Apis florea*	*Melipona irridipennis*
Common Name	Rock bee	Indian bee	Little bee	Dammer or stingless bee
Size	Very large	Medium	Very small	Very tiny
Habitat	Found in plains & hilly tracts	Found in plains & high altitudes	Found in plains	Found in plains
Comb	Large single vertical comb	Build parallel combs	Build single, small, vertical comb about the size of palm	Build sac like comb
Occurrence of comb	Comb is seen suspended from rocks, branches, tall trees, ceilings of neglected and uninhabited houses	Comb found in old buildings, forests, broken posts	Comb is suspended from branches of small trees, roof of buildings	Build nests in hollows and crevices of trees, rocks & walls
Habit	Migratory & Ferocious	More prone to swarming* & absconding**	Migratory, Do not sting	Do not sting but bite
Domestication	Wild, cannot domesticate.	Domesticated	Domesticated	Domesticated
Annual Honey Yield	36 Kg honey / comb / year.	6-8 kg / colony / year	½ Kg / year	Poor yield

* **Swarming** is the process of forming a new honey bee colony when the queen bee along with a group of workers leaves the colony and forms a new colony. In swarming a colony divides into two where one half remains in the old comb and the other half find a new home.

****Absconding** is the process when the entire colony of honey bees leaves its home in search of a new one. It can be due to shortage of nectar, extreme weather and climatic conditions and severe attack by predators and enemies.

Characteristic features of Queen, Worker and Drone

Characteristic	Queen	Worker	Drone
Size	Large	Small	Medium
Sex	Female	Female	Male
Head	Round	Large & Triangular	Triangular
Thorax	Broad & Deep	Small	Medium
Abdomen	Large & Pointed	Large & Pointed	Short & blunt
Sting	Curved	Straight	Absent
Wings	Appear shorter	Long	Long
Tongue and legs	Tongue smaller, legs are not modified. Wax glands are absent.	Tongue is well developed for sucking. Legs are highly modified and have pollen basket, pollen brush and wax glands are present.	Tongue & legs are not modified. Wax glands are absent.

Bee keeping Equipments

Bee Box

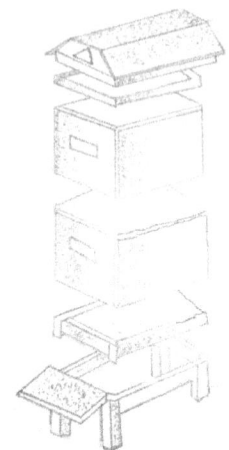

The most important equipment for bee keeping is the bee box. It consists of bottom board, brood chamber, super chamber or honey chamber, inner cover, top cover and frames. The bee box is placed on a stand. Bottom board forms the base of the box. Brood chamber is seen on top of the base and contains eggs and larvae and is followed by the super chamber which contains honey. The brood and the super chambers contain frames and the number of frames the chambers can hold depends on the size of the box. Above the super chamber is the inner cover and the top cover is on the top.

Honey Extractor

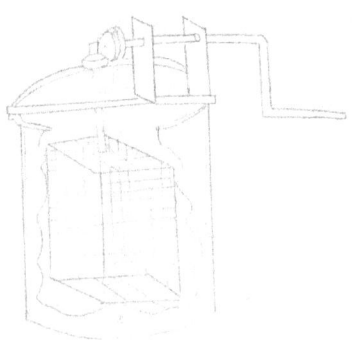

Honey extractor is used to extract honey in the purest form. The honey comb is whirled in a cage enclosed in an outside chamber. The honey is thrown out under centrifugal force. Hand operated and power operated extractors are available.

Queen Excluder

It is a device used to confine queen to certain parts of the hive. It can be a wire grid bound on metal or wood. The perforations are large enough for the worker to go through but small for the queen or drone to get through. These excluders are placed between the brood and super chambers.

Uncapping Knife

The bees seal the cells with wax when the cells are fully filled with honey. An uncapping knife is used to remove this wax cap for extracting honey.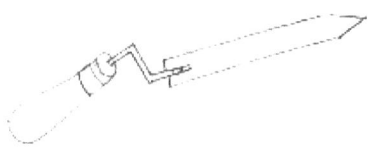

Queen Gate

It is a small metal plate with perforations. The holes are large enough for the worker to move but small for the queen. It is placed at the entrance of the hive while transferring a new colony to a bee hive.

Queen Cage

As the name suggests the queen cage is used for keeping the queen while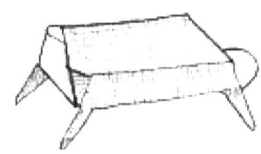

acquiring a new colony.

Smoker

Honey bees are sensitive to smoke and smoker contains a tin can with a spout for directing the smoke. Coir is used as the smoldering material to generate the smoke.

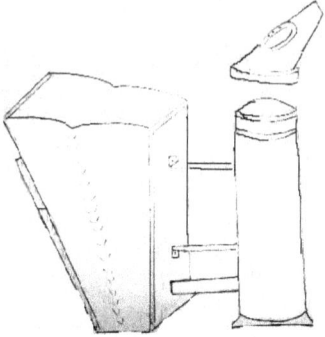

By products of Apiculture Industry

Beekeeping produces many by products like honey and wax, propolis and venom.

Honey is the most important product produced by bees from nectar by a process of regurgitation. It is sweet, easily digestible and source of energy. It is called as nature's antibiotic. It is used as sweetening agent, cosmetics, moisturizer, softener and in creams, soaps, shampoos and lipsticks.

Bee Wax is used in making candles, cosmetics, food processing, textiles, in varnishes and polishes, crayons, printing, in medicines, water proofing and for modeling and sculpture making.

Propolis is a resinous mixture of flavonoids, phenolics, and various aromatic compounds collected by the honeybee from plants. Propolis is used by bees for lining the inside of the nest and combs and to repair cracked hive. It is used in cosmetic making, medicine and in food technology.

Bee Venom is anti-inflammatory and used for arthritis. It is also used for treating sting hypersensitivity, central nervous system disorders. Apitherapy is the term given for the medical procedures using bee venom.

Simple methods for determining the purity of honey

Test	Procedure	Observation	Inference
Water Test	Take some water in a glass and add a drop of honey into it.	a. If it settles down b. If it dissolves	a. Pure honey b. Impure honey
Spot Test	Take a piece of filter paper and put a drop of honey onto it.	a. If it spreads slowly b. If it spreads slowly	a. Pure honey b. Impure honey
Flame Test	Take a match stick and dip it on the honey. Strike the match stick on the match box.	a. If match box lights easily & flame keeps burning b. If match box lights with difficulty & the flame does not keep burning	a. Pure honey b. Impure honey

REFERENCES

http://www.fao.org/fishery/culturedspecies

Jhingran, V.G. 1991. Fish and Fisheries of India. Hindustan Publishing Corporation (India), Delhi, India.

ABOUT THE AUTHOR

Revathy S is working as Assistant Professor , Department of Zoology in St. Xavier's College for Women, Aluva, Kerala, India

www.ingramcontent.com/pod-product-compliance
Lightning Source LLC
Chambersburg PA
CBHW070339190526
45169CB00005B/1964